Alan McKirdy has written many popular books and book chapters on geology and related topics and has helped to promote the study of environmental geology in Scotland. His other books with Birlinn include *Set in Stone: The Geology and Landscapes of Scotland* and *Land of Mountain and Flood*, which was nominated for the Saltire Research Book of the Year prize. *Northern Highlands – Landscapes in Stone* was long-listed for the Highland Book Prize. He is also the author of *James Hutton: The Founder of Modern Geology*. Before his retirement, Alan was Head of Knowledge and Information Management at Scottish Natural Heritage. Alan is now a freelance writer and has given many talks on Scottish geology and landscapes at book festivals and other events across the country.

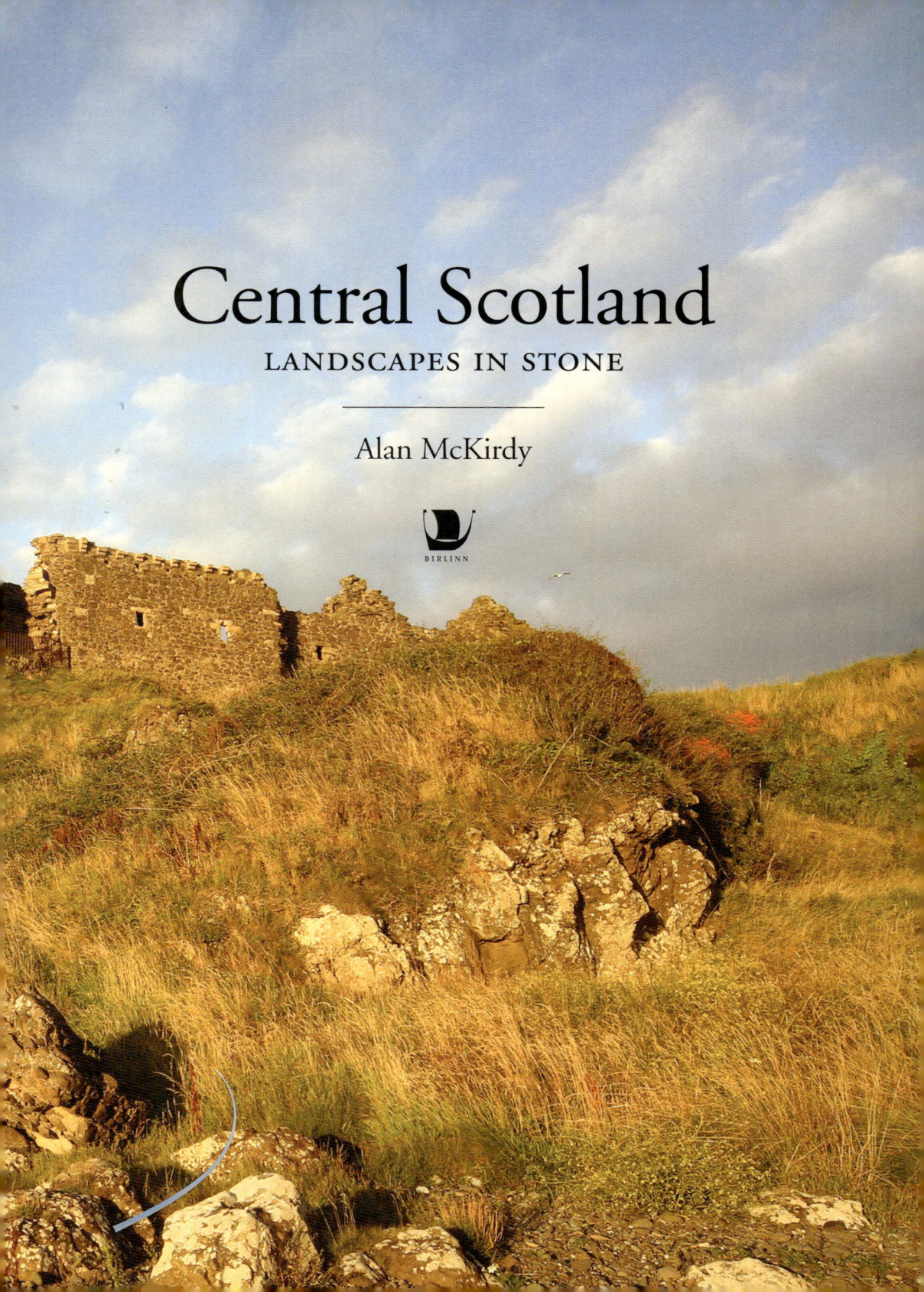

Central Scotland

LANDSCAPES IN STONE

Alan McKirdy

For Percy Alan Klar

First published in Great Britain in 2022 by
Birlinn Ltd
West Newington House
10 Newington Road
Edinburgh
EH9 1QS

www.birlinn.co.uk

ISBN: 978 178027 749 3

Copyright © Alan McKirdy 2022

The right of Alan McKirdy to be identified as the
author of this work has been asserted by him in accordance
with the Copyright, Designs and Patents Act, 1988

All rights reserved. No part of this publication may be
reproduced, stored, or transmitted in any form, or by any means,
electronic, mechanical or photocopying, recording or otherwise,
without the express written permission of the publisher.

Every effort has been made to trace copyright holders.
The publishers will be pleased to make good any omissions
if brought to their attention at the earliest opportunity.

British Library Cataloguing-in-Publication Data
A catalogue record for this book is available
on request from the British Library

Designed and typeset by Mark Blackadder

FRONTISPIECE:
Dunure Castle, Ayrshire

Birlinn Ltd would like to thank

for their generous donation towards this publication.

Printed and bound by Gutenberg Press Ltd, Malta

Contents

Introduction   7

Central Scotland through time   8

Geological map   10

1. Time and motion   11
2. Ancient inliers   14
3. Old Red Sandstone times   17
4. Life on Earth during Old Red Sandstone times   20
5. Carboniferous times – a verdant world   23
6. Desert storm   32
7. Towards the Ice Age   36
8. Golf – nature provides the perfect stage   41
9. Places to visit   43

Acknowledgements and picture credits   48

Introduction

The written history and archaeological records of Central Scotland only take us back a few thousand years, as far as the Roman occupation and Pictish times. The Earth's history, or the geology of the area, stretches back a further 400,000,000 years. For those who are unfamiliar with the subject, these are scary numbers and perhaps difficult to believe. The purpose of this book is to provide a guide that makes sense of the long and turbulent geological story that created the familiar landscapes and landmarks we see around us today across Central Scotland.

A rich pageant of geological history is laid out between our two principal cities, Edinburgh and Glasgow. The mineral wealth that lies beneath the ground, namely coal and iron ore, provided the raw materials that drove the Industrial Revolution in Scotland. The early focus on understanding the rocks beneath our feet was unsurprisingly initially centred on the most economically valuable mineral resources. Over time, scientific curiosity drove a wider investigation of the geology, and the full range of strata and rock types was gradually revealed.

The Geological Survey of Great Britain had an early outpost in Edinburgh. Its surveyors produced maps and accompanying memoirs, documenting the geology of Central Scotland.

The oldest rocks are found near Lesmahagow and in the Pentland Hills. They are known geologically as 'inliers': small areas of rocks from an older age surrounded by younger strata. Rocks of the Old Red Sandstone and the succeeding Carboniferous era underlie the rest of Central Scotland in almost equal measure. Explosive volcanic rocks, thick sequences of passively erupted lavas, desert sandstones, limestones and coal-bearing strata make up this bedrock patchwork. Some 250 million years later, the Ice Age came. A covering of ice and snow up to two kilometres thick blanketed the landscape for a further two million years. Glaciers sandpapered and burnished the bedrock, as they slowly ground their way from higher ground to lower altitudes, producing the familiar scenes and landmarks we see today.

Opposite.
Dumbarton Rock.

Central Scotland through time

Period of geological time	Millions of years ago	Scotland's global position	Environments and events in Central Scotland
Anthropocene	Last 10,000 years	57° N	• Between the sixteenth and nineteenth century, the Little Ice Age caused a significant lowering of temperatures. • Coal-mining started more than 200 years ago.
Quaternary	Started 2 million years ago	Present position of 57° N	• 9,000 to 6,500 – sea levels rose and raised beaches were formed. • 11,500 onwards – the ice retreated as the climate started to warm. • 12,500 to 11,500 years ago – the climate became very cold as the ice returned; the Loch Lomond Readvance. • 14,700 to 12,500 years ago – for a brief interlude, temperatures were similar to those of today. • 29,000 to 14,700 years ago – the landscape was entirely covered by an ice sheet during this, the last advance of the ice. • Before 29,000 years ago and for a period approaching the last 2 million years, there were prolonged periods when thick sheets of ice covered the area.
Neogene	23–2	55° N	Temperatures fell as the Ice Age approached.
Palaeogene	66–23	50° N	Between 65 and 60 million years ago, the ancient continent of Pangaea was split asunder and the North Atlantic Ocean formed.
Cretaceous	145–66	40° N	Sea levels rose to drown the area, but no rocks of this age are preserved Central Scotland.

Period of geological time	Millions of years ago	Scotland's global position	Environments and events in Central Scotland
Jurassic	201–145	35° N	Skye was our 'Jurassic Park', but no rocks of this age are preserved in Central Scotland.
Triassic	252–201	30° N	Desert conditions with ephemeral river systems prevailed across 'Scotland', although there are no deposits of this age in Central Scotland.
Permian	299–252	20° N	Deserts covered the land. Evidence for these dunes is to be found near Mauchline.
Carboniferous	359–299	On the Equator	'Scotland' was located at the Equator and tropical rainforests were widespread, as was volcanic activity.
Devonian	419–359	10° S	Arid conditions with extensive river systems were widespread across the Old Red Sandstone continent, as was volcanic activity. Primitive insect life, plants and fish inhabited this barren world.
Silurian	444–419	15° S	Large upheavals created the Highlands of Scotland and the Southern Uplands as the Iapetus Ocean closed.
Ordovician	485–444	20° S	'Scotland' was located on the southern shores of the Iapetus Ocean.
Cambrian	541–485	30° S	What are now 'Scotland' and 'England' were separated by the width of the Iapetus Ocean.
Proterozoic	2,500–541	Close to the South Pole	No rocks of this age are preserved in Central Scotland.
Archaean	Prior to 2,500	Possibly close to the South Pole	The age of the Earth is around 4,543 million years.

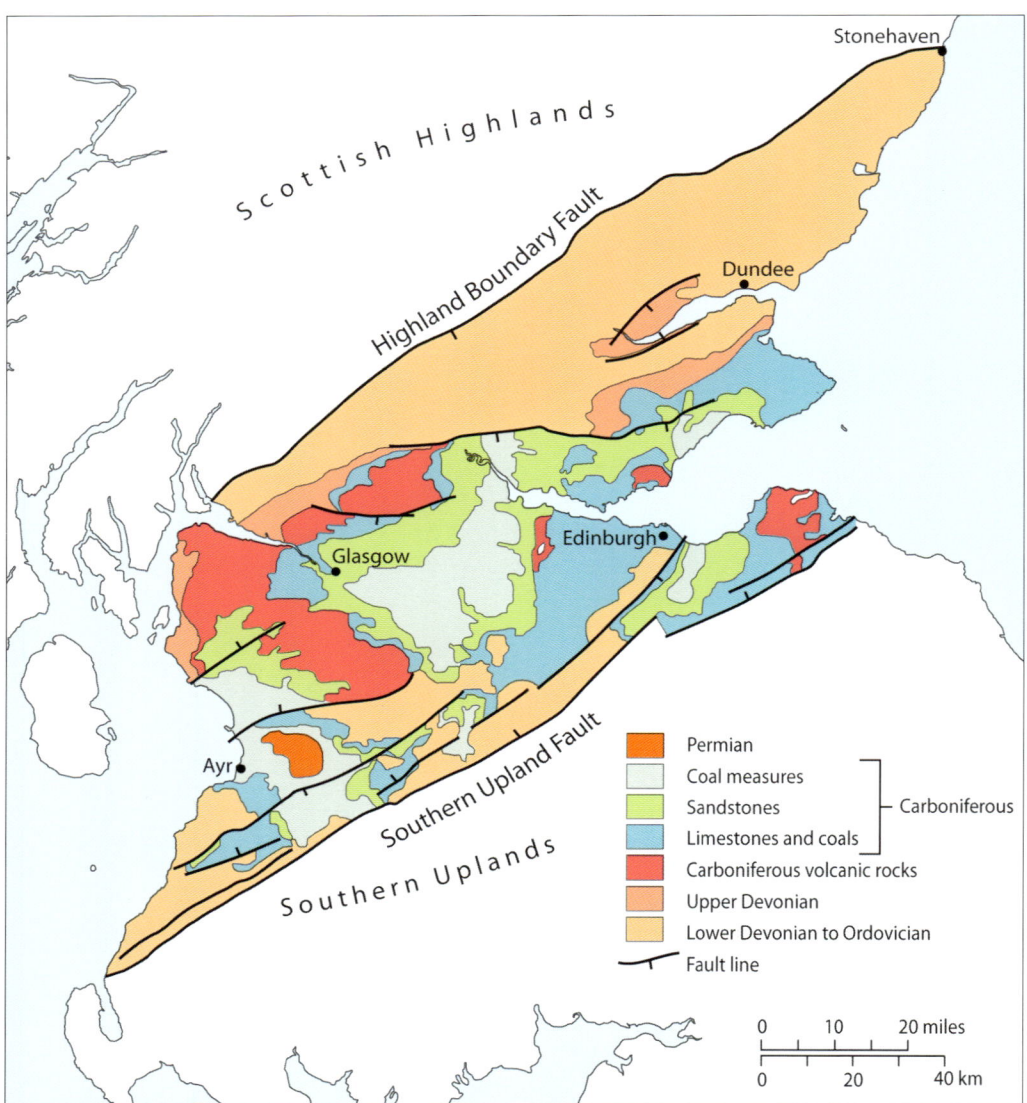

Central Scotland is defined by two great tears or faults in the Earth's crust. To the north, the Highland Boundary Fault separates the softer contours of the highly populated Central Belt from the rugged landscapes of the Grampian Highlands. Defining the southern boundary is the Southern Upland Fault, with the pillow-like hills of the Moorfoot and Lammermuir Hills lying beyond. Between these prominent landscape-scale features lie undulating lowlands underlain by strata of predominantly Devonian and Carboniferous age. As we'll hear later, Scotland was located on or close to the Equator during these geological periods, so these strata represent a variety of different environments including desert, river plains and tropical rainforest. This was an unstable landscape. There is extensive evidence of volcanic activities, with great volumes of lava passively disgorged and also explosive eruptions of ash and pyroclastic flows. Desert conditions were reasserted during Permian times and thick layers of sands are preserved around the town of Mauchline. One of the final events to leave a mark on the area was a series of thin ribbons of molten rock that cut through the ancient landscape as the Mull volcano blew its top.

1
Time and motion

Time

The Earth was formed around 4,543,000,000 years ago from swirling clouds of space dust and rocks. Much of this long and fascinating history of the planet is preserved in the record of the rocks on which we walk. But it only makes sense if you can read the clues. Geologists are like detectives, interpreting the signs of past environments and ecosystems made up of long-extinct plants and animals. Layers of rock are like the pages of the Earth's autobiography and can be read to piece together this incredible story. Central Scotland reveals many fascinating glimpses of key episodes in this geological narrative and is also the place where many important scientific discoveries were made.

Understanding time is crucial to the appreciation of this geological story. For millennia, the timescale on which the Earth was formed followed a literal interpretation of Genesis. The heavens and the Earth were created in six days, with a day of rest thrown in for good measure. This was accepted as an unassailable fact until an Edinburgh man, Dr James Hutton, revealed the truth of it. Around 250 years ago, during that golden age of scientific enquiry known as the Scottish Enlightenment, Hutton studied the rocks of his native land within 200 miles from his home and came up with a greatly extended timescale. He could see 'no vestige of a beginning and no prospect of an end' to the time required to shape the natural landscape.

It was only during more recent times that an adopted Scot, Arthur Holmes, Regius Professor at Edinburgh University, was able to quantify the timescales involved. Instead of days and weeks, the age of the Earth and the complex jigsaw of rocks that made up its surface should more accurately be measured in millions of years.

Geologists have divided the yawning stretches of geological time into manageable chunks known as 'geological periods'. Each is defined by a specific span of years, with a date-stamped and a universally agreed beginning and end, although in true scientific fashion these

Dr James Hutton (1726–97) is recognised worldwide as the founder of Modern Geology. The ideas in his ground-breaking book *Theory of the Earth* were not at first universally accepted, but they have stood the test of time and still underpin the modern science of geology.

The plates are propelled across the globe by heat that comes from the Earth's core. The temperature at the centre of the planet is around 6,000°C and inevitably heat leaks to the adjacent layer, known as the mantle. This sets up a convection motion below the Earth's crust that moves the plates. This heat source shows no immediate signs of cooling, so this force will ensure that continents continue their restless journeys across the face of the Earth. This process of continental drift happens at such a slow rate that a human lifetime is insufficient to observe much change. The geography we recognise today will look very different in millions of years as oceans widen, continents split and landmasses collide. This pattern of constant change is clearly evident in the geological record and will continue long into the future.

dates are continually refined. This standardised approach has helped scientists to correlate rocks, fossils and significant events of similar age across the country and indeed around the world. Pages 8 and 9 of this book place the events that have left a mark on Central Scotland in date order. This sequence is known as a geological column. The geological periods are arranged from the oldest to the youngest at the top. We are still in the Anthropocene Period, the most recent and the only period of geological history where our species, *Homo sapiens*, has played a part, and not always a positive role, in affecting the atmosphere, geosphere, biosphere and hydrosphere of Planet Earth.

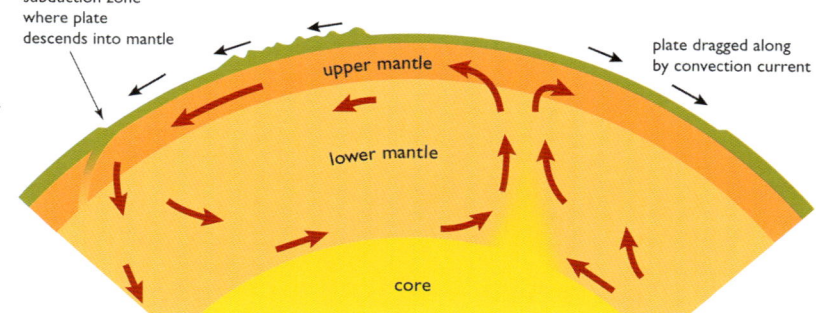

Motion

Another unfamiliar concept is that the rocks on which we stand are in constant motion and have been since the formation of the Earth's crust more than 3,000 million years ago. The Earth's surface, the crust, is divided up into seven large chunks, or plates, and more than a dozen smaller ones that move independently of each other. Plate tectonics is our model of how the surface of our planet behaves. It has only been understood since the middle of the last century, but it explains much of the function of the Earth's crust today and, equally importantly, how it worked in the geological past. Each tectonic plate moves at an average rate of around 6 cm each year. Doesn't sound like much, but over millions of years continents have been shunted from one side of the globe to another. The land that would become Scotland came into existence near the South Pole and has travelled relentlessly northwards through every climatic zone that the planet has to offer. It's been held in the deep freeze of an ice age and then experienced the scorching heat of desert conditions, as it was propelled into equatorial latitudes. All of these changes are faithfully recorded in our varied, albeit incomplete and sometimes difficult to interpret, record of the rocks.

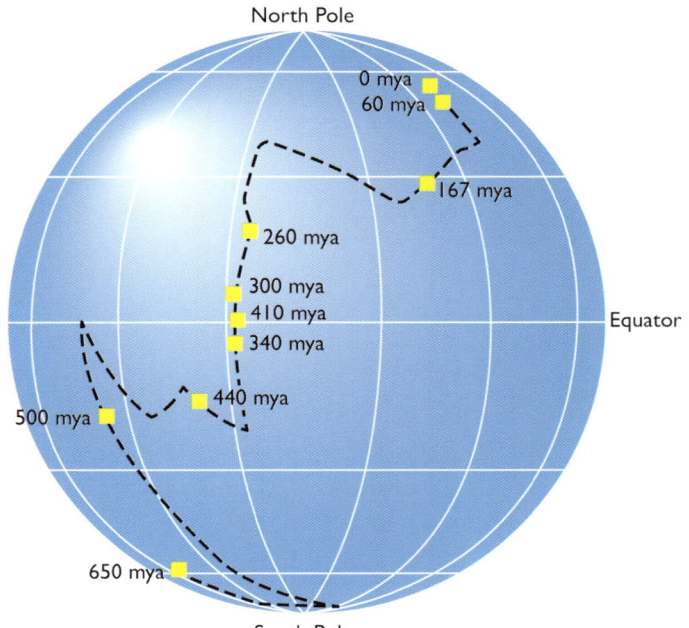

We can chart Scotland's journey across the globe over the last 650 million years. Starting near the South Pole, the land that would become Scotland has moved relentlessly northwards since that time. This chunk of the Earth's crust has moved through every climate zone, from freezing to scorching in the desert sun. This helped to determine the strata that makes up our bedrock and the type of plants and animals that inhabited the transient ecosystems that are represented in the geological record.

2
Ancient inliers

Near Lesmahagow, south of Glasgow, ancient rocks of Silurian age peep through a cover of younger rocks. This window into past times is made from rocks with strong Southern Uplands affinities – ancient muds, sandstones and siltstones. These strata are fossil rich, containing a variety of mussel-like bivalves, ancient and long-extinct arthropods known as trilobites, and larger creatures called eurypterids or sea scorpions that preyed on small fish and arthropods. The most important find was that of one of the most primitive fish known from anywhere in the world. The jawless creature, known as *Jamoytius*, was

Jamoytius kerwoodi is related to the modern-day lamprey. It has a sucker for a mouth and a row of gill openings down its body. After specimens of this species were described by Traquair, the Glasgow Geological Society set up an annual summer camp near the site and intensively collected from the area. Gunpowder was used to loosen the rock. The site is now protected and further collecting is restricted.

first recorded by one of Scotland's most renowned fossil fish experts, Dr Ramsay Heatley Traquair (1840–1912). Many of the leading lights of Scottish geology have also visited this place and published their findings in learned journals. Jawless fish are rare worldwide and *Jamoytius* is important to evolutionary studies and also in relating these rocks to others of the same age found in far-flung parts of the world, including Spitsbergen, Canada and Russia.

Inliers of older rock also occur in the Hagshaw Hills; the North Esk inlier, which is part of the Pentland Hills; around Girvan; and, finally, south of Stonehaven. Most of these rock sequences are fossil bearing and give an invaluable insight into life on Earth during these ancient times.

The Girvan inlier provides key evidence about one of the most significant events that shaped Scotland. Unlikely though it may seem, Scotland and England were at one time separated by an ocean wider than the North Atlantic Ocean is today. As the tectonics plates rearranged themselves, so the continental landmasses carrying Scotland and England converged and the intervening ocean, known as the Iapetus Ocean, disappeared.

This beautiful fossil trilobite, *Encrinurus pagei*, was unearthed from the Silurian-age inlier just to the south of Edinburgh, near Carlops. It is known as the strawberry-headed trilobite on account of the bizarre pattern of bumps that cover its head. It is rare to find such a perfectly preserved specimen. In life, this creature inhabited a shallow sea, burrowing through the sands and muds on the sea floor.

Layers of sand and mud built up on the Iapatus Ocean floor. As continents collided, what we now recognise as the Southern Uplands was created. The Highland Boundary Fault and Southern Upland Fault became defining features for Central Scotland. The Girvan inlier is a surviving fragment of the ocean floor that was scraped off the descending plate as the ocean closed. Such occurrences are extremely rare in the geological record; where they are recognised, they become the focus for intensive study. These fragments are known as 'ophiolites' and have also been recognised at Arrochymore Point in Loch Lomond and Garron Point, north of Stonehaven.

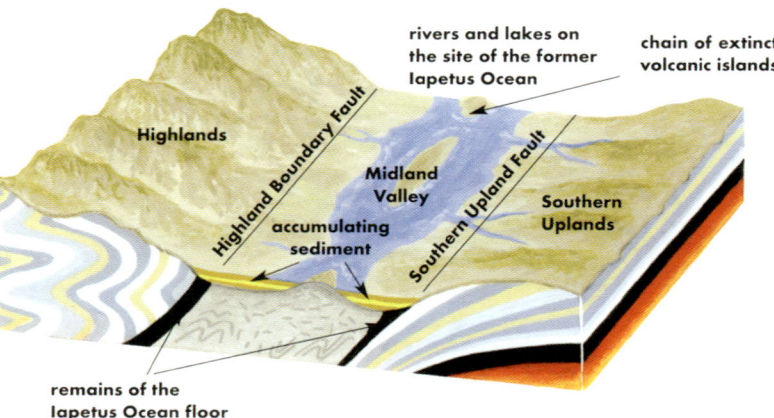

This sequence of lava flows, found to the south of Girvan, are known as pillow lavas. They are characteristic of molten rocks that were erupted underwater onto the floor of the ancient and now disappeared Iapetus Ocean. The skin of each pillow of lava cooled almost immediately and they stacked one on top of another as the eruption continued.

3
Old Red Sandstone times

Geologists have called the new land created by this dramatic continental rearrangement 420 million years ago the Old Red Sandstone continent. It is also known as Laurussia.

The Old Red Sandstone continent was a raw and hostile place to the early primitive plants and animals that established a presence here. Volcanic eruptions, frequent earthquakes and the searing heat of this arid landscape all combined to make this place a hell on Earth. At this time, around 400 million years ago, 'Scotland' lay 15° south of the Equator in an unlikely alliance with North America and Greenland. Despite its global position so close to the Equator, great rivers on the scale of the Ganges and Mississippi flowed across the otherwise parched plains. The largest of these rivers flowed south-westwards, joined by tributaries shed from the Grampian Mountains to the north.

Continued movement along the Highland Boundary Fault helped

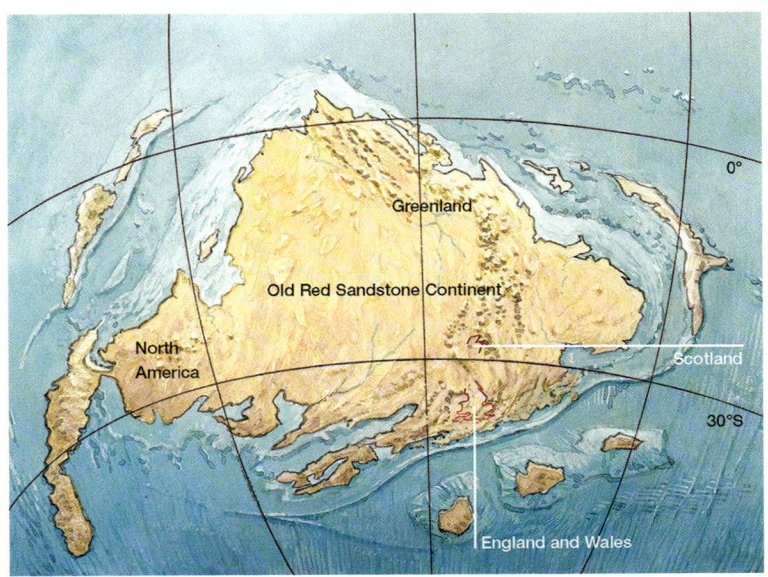

The Old Red Sandstone continent was an unlikely amalgamation of bedfellows. This land was the result of continental collisions that created new and towering mountain ranges of Himalayan heights.

Around 400 million years ago, the landscape was arid and inhospitable. The newly formed mountain ranges were rapidly eroded by natural forces. Rivers and streams carried the debris of erosion to the lower ground, where it was dumped into piles sorted by the river currents. Primitive plants, the forerunner of all greenery we see around us today, had just evolved and were beginning to get a toehold in this hostile world.

to create a depression in the Earth's crust, which became flooded by streams and rivers that flowed from the higher ground to the north. The rivers that emerged from the rapidly eroding mountains into this basin were loaded with sediments. Boulders, pebbles, sands and muds carried by the rivers built up into layers of sandstones and conglomerates. In total, it is estimated that a thickness of around 8 km of sediments accumulated at this time. These layers were rusty red when they were formed and retain that colouration to this day. It is as a result of this characteristic that the Old Red Sandstone was thus named.

These red sandstones are characteristic of the Old Red Sandstone. Layer upon layer of sand, with occasional pebbles, were laid down on the beds of fast-flowing rivers. These sediments were banked up into small dune forms and ripples on the riverbed. Careful study can reveal the direction in which the rivers flowed.

Great thicknesses of conglomerate, also known as 'puddingstone', also built up in layers. The pebbles are rounded, which indicates they have been transported great distances by the rivers. They tumbled over each other many times in this high energy environment and were only deposited in layers when the river current slackened sufficiently. It is revealing to study the nature of the pebbles; they can help us pinpoint the nature of the bedrock over which the rivers flowed and therefore their direction of flow.

Dunnottar Castle sits on one of the most impressive occurrences of conglomerate to be found anywhere in Central Scotland. The largest well-rounded boulders are up to a metre in size. This gives an indication of the scale of the river that must have existed here around 400 million years ago. Many boulders and pebbles are comprised of quartzite, so the journey would not have been a short one to knock the corners off such a tough material. Exposure of these deposits continues in the cliffs extending southwards towards Tremuda Bay. This cliff section to the south also exposes a thick sequence of volcanic rock that was erupted over the conglomerates. Younger beds of conglomerate, in turn, overlie these basalt lavas. The juxtaposition of these ancient deposits paints a picture of a high energy and unstable environment, with fast-flowing rivers and active volcanoes.

4
Life on Earth during Old Red Sandstone times

Plants

The landscape during Old Red Sandstone times was barren and almost bereft of plants. Greenery of any kind was rare at this time anywhere on the planet, as it was very early in the evolutionary development of plants. Those that did exist were primitive. The fossil record is fragmentary and incomplete, but it does provide an insight into the early development of members of the plant kingdom. One of the most enigmatic is the moss known as *Parka* that grew as a thin sheet over the rock surfaces swept clean after river-flooding events.

The moss, known as *Parka*, was described by the famous fossil collector Hugh Miller and by other eminent scientists of the nineteenth century. Although it still defies a definite taxonomic classification, it is most likely to be a primitive moss. The best specimen was recovered from a site near Forfar.

Fish

This arid world, criss-crossed by rivers with abundant pools and lakes, provided many places for primitive fish to thrive. The diversity of these primitive animals is breathtaking. Their fossilised remains are found across Central Scotland in red sandstone deposits, most frequently in finely laminated layers that are indicative of quiet lagoon or lake conditions. They may have lived in higher energy environments, but their remains would have been broken up and dispersed. These fossils have attracted the attention of scientists from Victorian times to the present day. During this prolonged period of study, around 30 different species of early fish have been recovered from the Old Red Sandstone. They are recognisable as fish but are varied in their form and biology. Some have jaws, whilst others do not. Many species have spines, probably evolved for defence against predators.

One of the most extraordinary fossil finds was made at Dura Den in Fife. A distinguished party, including Lord and Lady Kinnaird and the celebrated geologist Roderick Murchison, visited the site in 1858. During their time there, a three-foot-long specimen was uncovered of *Holoptychius*, a bony fish with similar characteristics to present-day species. Later, more of the quarry was cleared and 'nearly a thousand fish were lifted from their stony bed of ages'. An assemblage of fish carcasses was also discovered, which led observers to conclude that they had all perished at the same time as the lake filled with blown sand. Some of the fish had ingested sand, confirming the theory. This was a famous discovery, and rock slabs exhibiting grisly scenes of mass fish mortality are now held by many museums, including the Natural History Museum in London.

Volcanoes and lavas

Volcanoes and lava flows were a significant part of the Old Red Sandstone story. This unstable landscape was sliced by faults and shaken by earthquakes. Lava flows periodically scarified the land. Some of the oldest eruptions of this age are found in the Ochil Hills, where lavas and pyroclastic deposits, similar in nature to those that destroyed Pompeii, are common. The lavas were spewed onto the land surface through fissures in the Earth's crust, some of which tapped a deep source.

These lavas extend north-east and seamlessly become the Sidlaw Hills. Thin lenses of lava are interleaved with the predominant sandstone layers. These strata continue to the coast around Crawton and Montrose.

View of the Ochil Hills, showing the Old Red Sandstone lavas rising prominently from the Strathearn Valley below. The hills were resistant to erosion, whereas the sandstones that underlie the valley floor were planed flat by the passage of ice in later times.

5
Carboniferous times – a verdant world

Scotland continued its meandering path northwards, approaching the Equator around 360 million years ago. Deserts were replaced by tropical conditions, accompanied by a great flowering of vegetation. Rather than an arid landscape with a sprinkling of primitive plants, as in Old Red Sandstone times, Carboniferous Scotland boasted lush tropical rainforests. There were insects such as dragonflies, with a wingspan of a metre or more, and azure-coloured seas inhabited by sharks, corals and a wide variety of fish species. Biodiversity flourished in equatorial latitudes, as forests spread over huge areas of the globe. Amphibians and reptiles made their presence felt and insect life diversified. Part of the reason for this explosion of life may have been the increased oxygen levels in the atmosphere. However, as biodiversity blossomed, there were also some serious volcanic eruptions to contend with.

Fire and brimstone

Around 355 million years ago, thick sequences of lava were belched from fissures and volcanic vents across Central Scotland. These lava flows, now long cooled, are still very prominent landscape features. The Touch, Fintry and the Gargunnock Hills, the Campsie Fells, the Renfrewshire Hills, the Eaglesham Uplands and the Bathgate Hills are just a few of the piles of lava that were erupted during early Carboniferous times. In some instances, the lavas were erupted onto flat coastal plains, with occasional lagoons. In such circumstances, there were explosive interactions between the water-saturated sediments and the molten rock.

Glasgow is encircled on three sides by this lava field. The lavas are not of uniform composition, and over time changed subtly in terms of their chemistry and mineral content. Basalt and the slight variant (and exotically named) hawaiite predominate. What is remarkable here is the accumulated thickness of the erupted lavas. In places, these stacked pancake layers reach a thickness of over 800 m. Few prominent volcanic vents have been identified, so these lavas fields are thought to have flowed from fissure eruptions like this example from Iceland.

Although the lavas were predominantly erupted through fissures, there are places where more conventional volcanoes are present.

Right.
This is a modern-day fissure eruption in Iceland, which illustrates the manner in which the rocks of the Gargunnock Hills and Campsie Fells were erupted.

Below.
Successive layers of lava were extruded to create this landscape at Duntocher, near the Erskine Bridge. This process was repeated time after time to create the step-like appearance of these hills.

Dumbarton Rock is one such example. The imposing defensive site has been fortified since the fifth century and occupies a dominant position overlooking the Clyde Estuary. It is circular in plan and has punched its way to the surface through older sedimentary strata. Lumps of these sandstones were stripped away as the magma ascended through the Earth's crust and they are now isolated blocks incorporated into the lava. Dumbarton Rock would have been part of a much larger volcanic structure that has been diminished by later erosion. What remains is a central plug of basalt lava that cooled in the lower reaches of this structure, only to see the light of day after the upper part had been removed.

Slightly later in its timing was the intrusion of what is known as the Midland Valley Sill – similar in scale to the Whin Sill that traverses northern England. A thick sheet of molten magma was injected into existing sedimentary layers from a deep source. Much of it still lies

Loanhead Quarry near Beath gives a clear indication of the way in which the lavas accumulated during this period. Each flow is separated from the preceding and succeeding one by a thin, red fossil-soil horizon. There is also a clear indication that considerable periods of time elapsed in which soils had time to develop, supporting an associated covering of vegetation before the next eruption event.

Dumbarton Rock is the remains of a volcano that was active around 300 million years ago.

Stirling Castle sits on rocks that form part of the northern edge of the Midland Valley Sill. These impressive ramparts are made from dolerite, a coarser-grained form of basalt. This magma didn't quite make it to the surface and cooled more slowly at depth. The strata that once lay above the sill have been removed by later erosion.

underground, but its northern and southern edges are prominent landscape features in Central Scotland. The Lomond Hills are capped by this pulse of hot rock, as are the ramparts on which Stirling Castle is built. Those who have used the old Forth Road Bridge may have noticed impressive road cuttings through the Midland Valley Sill on the north side of the Forth. This rock is an ideal road stone, so it has been worked extensively at a number of locations for that purpose.

Tropical seas and coral reefs

Central Scotland was akin to the modern-day Bahamas around 325 million years ago. The volcanoes described above were active at the time, and lagoons, tropical seas and coral reefs fringed the coastline as it existed then.

There were dramatic variations in sea level at this time and incursions and retreats by the sea were frequent. Some of the recently erupted lava landscapes were periodically submerged under the waves. Layers of limestone formed on the sea floor, as a mixture of chemical precipitate and broken shells built up.

Lagoons also developed along the coastline, close to the open sea. Rivers flowing across the adjacent landscape introduced sands and muds to the coastal environment. These flowing waters deposited their load of sediment as they reached the sea, producing advancing deltas. New land built up at the coast as a result. Tropical forests rapidly developed on this newly emergent swampy ground and they thrived until the next incursion by the sea.

This cycle of growth and flood was repeated many times, driven by sea-level change. These environmental changes are preserved in the record of the rocks, represented by deposits of limestone, sandstone, ancient soils and coal. Concentrations of ironstones, later exploited during the Industrial Revolution, also accumulated on the floor of the shallow-water lagoons.

Further offshore, in quieter, undisturbed waters, thick layers of limestone built up. In places, reefs emerged from the sea floor, with some reaching 15 metres in height. Corals and sea lilies (also known as crinoids), sponges, squid and oysters thrived in these sheltered waters.

Sharks' teeth and an almost full skeleton of that animal have also been recovered from rocks of this age. This unexpected find was made in Bearsden and was given the scientific name of *Akmonistion zangerli*. Danger lurked even in these tranquil waters.

The consequences of rapid environmental change are neatly captured in this diagram, showing how different conditions give rise to a wide variety of rocks, from sandstone to coal and ironstone.

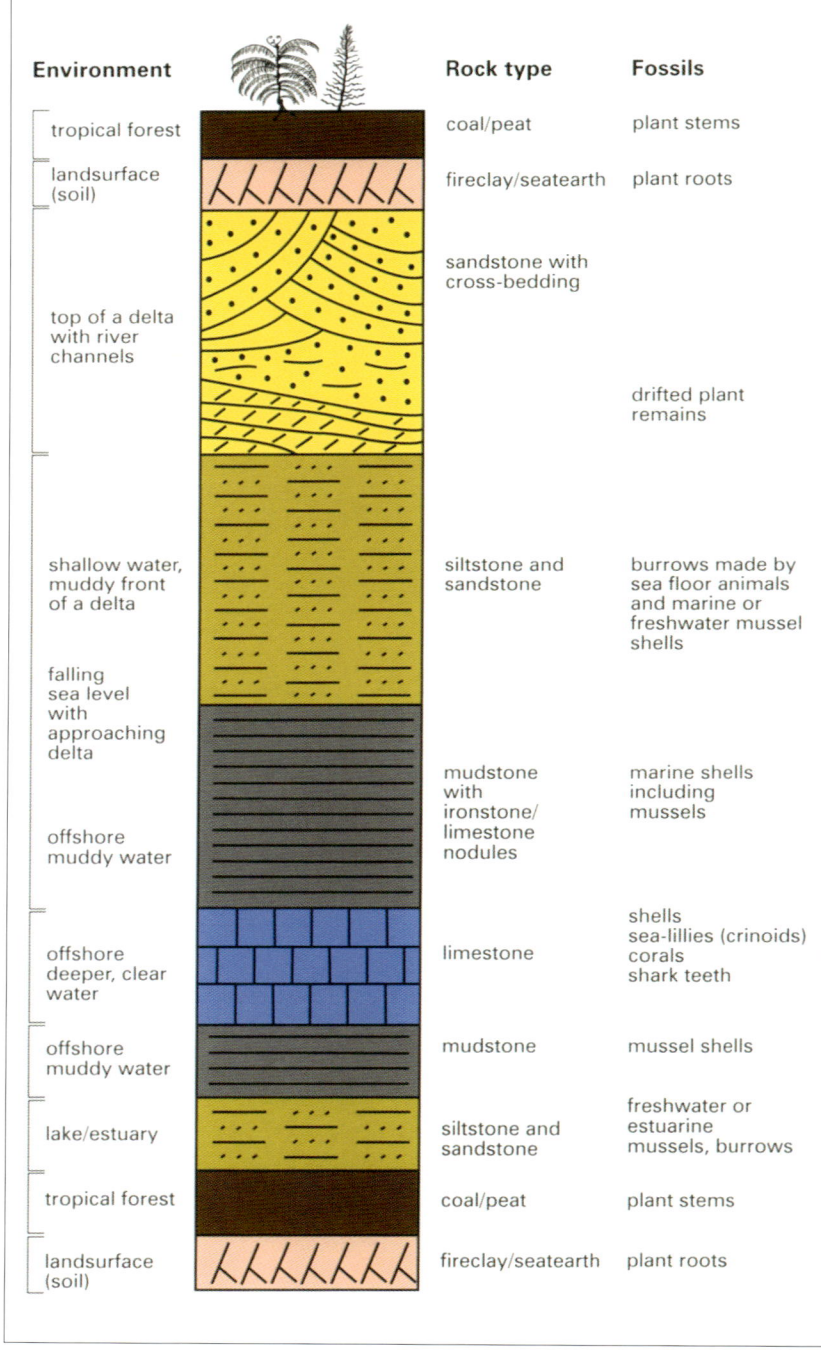

Fossil Grove in Glasgow is an amazing surviving remnant of the coal swamps that covered the country around 325 million years ago. It is significant for the eleven tree roots that are on display and also as one of the first acts of geological conservation to have taken place anywhere in the world. These stumps were uncovered in the course of quarrying activities in Victorian times. Their heritage value was quickly realised and a building was constructed to save them from the elements. The building has been refurbished since then and it is still in the care of the local authority.

This is a reconstruction of what the Bearsden shark would have looked like in life and the other animals with which it would have co-existed. Ray-finned fish and shrimp-like crustaceans lived alongside this top predator. The preservation of this specimen is excellent. Muscles and blood vessels are well preserved. The large structure attached to the back of its head may have been part of a mating ritual.

The age of coal

Later in Carboniferous times, dense swathes of tropical rainforest once again covered much of the land. This was to become another age of coal. The formerly productive coalfields we see today are limited to Ayrshire, Fife, south of Edinburgh and a larger area running between Alloa and Motherwell. The extent of the ancient tropical forests would have covered the greater part of Central Scotland. These natural deposits of coal and associated deposits of iron ore were to support the development of heavy industry in Central Scotland and a workforce of thousands of men and women. Few geological eras have had such a profound effect on the social and economic fortunes of the country. The age of coal is now over in Scotland, with the realisation that burning this fossil fuel contributes so adversely to global warming. The pitheads and winding gear are now silent and a way of life has completely disappeared. It should be remembered that, for centuries, this black gold was the county's economic foundation.

Let's rewind to the middle years of the Carboniferous Period, 310 million years ago, when the tropical rainforests were at their zenith. The pattern of rapidly changing sea levels persisted from earlier times. Established forests were drowned by the rising tide. Sands and muds were then dumped on top of the decaying organic remains of the swamped rainforests by the higher sea levels. This pattern was repeated time without number, so cycles of deposition and compaction, similar to that illustrated on p. 28, were established across the country. As the sands weighed down on the rotting organic matter provided by the forest, so layers of coal were created by the heat and pressure of this deep burial.

One of the final acts of the Carboniferous Period can be seen on the Fife coast beside St Monans harbour at low tide. Shock waves from continental collisions in Europe travelled north to buckle these Carboniferous rocks. Most sedimentary rocks are deposited flat, but these strata have been upended and tightly folded, illustrating the colossal forces that must have been at work.

Left.
This reconstruction of the tropical coal swamp shows the vegetation that clothed the landscape. Insects also inhabited this world. Dragonflies grew to enormous sizes, as oxygen levels reached 28 per cent of the atmosphere, considerably more than today. Wildfires were more frequent, too, the charred remains of vegetation providing the evidence.

Below.
At St Monan's on the Fife coast, Carboniferous-age strata have been bent into a tight fold.

6
Desert storm

During Permian times, Scotland continued to drift northwards from equatorial regions to the latitude that is today occupied by the Sahara Desert. The conditions were right for the development of an extensive desert landscape.

Only a small patch of sedimentary rock of Permian age is present in Central Scotland today, located near the town of Mauchline in Ayrshire. The oldest beds are lavas and thin pyroclastic flows, providing clear evidence of an initial explosive volcanic episode, albeit limited in geographic extent. Lying above are substantial wind-driven sand dunes that were later cemented by the percolation of iron-rich fluids to form solid rock. These rocks have been extensively quarried in past years, with many buildings in the west of Scotland constructed using stone quarried from the now disused Ballochmyle stone quarry near Mauchline.

The Fife coast is dotted with a dozen or more small volcanic vents

At the beginning of the Permian Period, 299 million years ago, Scotland was completely landlocked in the Pangaea supercontinent. Pangaea means 'all Earth', with all the land areas of the Earth locked in a lasting embrace.

that were active in late Carboniferous and Permian times. Coastal erosion has carved into the heart of some of these vents and careful study has revealed a great deal about how these volcanoes functioned. The volcanic vents punched through wet ground on their passage to the surface. Hot magma and ash reacted in an explosive manner to the saturated layers and surface water.

Several of the vents contain rocks that were brought to the surface from great depths in the Earth's crust and upper mantle. These fragments, in this case from a deep source, are known as xenoliths. They were ripped by the ascending magma from rocks some 70 km beneath our feet. Much research has been done on the mineral content of these xenoliths and inferences have been drawn about the nature of the Earth's upper mantle and lower crust as it existed some 300 million years ago. Samples of material of this age and from this depth are exceptionally rare.

This would have been the scene 250 million years ago in this part of Central Scotland. Pebbly beds containing the sparse remains of plants indicate the presence of occasional wetter episodes and perhaps flash floods.

These rocks are typical of the Mauchline sandstones for this period, pictured near Howford Bridge.

Above.
The Rock and Spindle, near St Andrews, is an example of a late Carboniferous or early Permian volcanic vent.

Left.
Xenoliths brought from deep in the Earth's crust and upper mantle are found at Elie Ness on the Fife coast.

7
Towards the Ice Age

Scotland had an eventful journey to its current position of 57 degrees north of the Equator, and also through geological time. In terms of our passage northwards, the great supercontinent Pangaea eventually broke up as Europe and North America charted different courses. The North Atlantic Ocean opened around 65 million years ago and Scotland became part of the European continental landmass.

Our journey through geological time was equally remarkable for how few reminders there are of the next 260 million years. After the desert sands and volcanoes of the Permian, the next period to leave its stamp on the landscape of Central Scotland was the Quaternary – better known as the Ice Age. The Triassic, Jurassic, Cretaceous, Palaeogene and Neogene Periods had come and gone and had left little impression here.

As Scotland travelled further north, temperatures fell. But larger forces were also at play. Planet Earth's orbit around the sun varies from elliptical to circular. It takes around 100,000 years for this cycle to be completed. When the orbit around the sun is circular, the climate is benign, but as the planet travels further from the sun during the elliptical phase, temperatures fall like a stone and glacial conditions prevail. This has happened many times during the last 2.6 million years. During periods of most extreme cold, sea levels drop dramatically, as much of the water on the face of the Earth is locked up as ice. Coastlines change out of all recognition. During the last cold phase, the North Sea was completely frozen over between Scotland and Scandinavia and a land bridge emerged that linked England to Europe. The most dramatic consequence of this change is that the landscape was, at times, entirely buried by ice. The ice sheet waxed and waned in response to global temperatures, but it was a frozen scene for long periods. Between the periods of maximum expansions of the ice, there were warmer times known as inter-glacials. We are in an inter-glacial period at the minute, but the ice may well return.

The effect of the ice on the landscape of Scotland was dramatic

TOWARDS THE ICE AGE 37

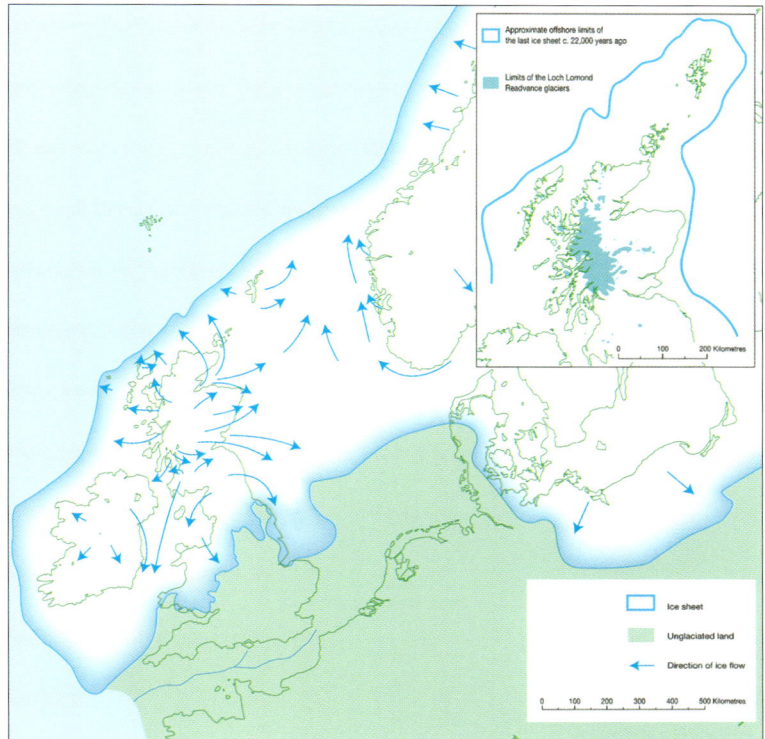

During the last advance, the landscape was entirely submerged by the ice. Early members of *Homo sapiens* (us!) who lived at this time walked freely across the English Channel. Large communities lived on land that is now at the bottom of the North Sea. These areas were later flooded, as temperatures rose and the ice melted.

and long-lasting. It shaped some of our most iconic landmarks and scenic vistas. The form of Dumbarton Rock, the plinth on which Stirling Castle sits, the Ochil Hills and countless other places that define us as a nation were formed as a result of the sandpapering the bedrock received as glaciers ground and scratched their way from high ground to lower altitudes. And when the ice melted, another set of landforms were created. Meltwater channels, kames and eskers date from these later glacial times, when the melting ice sheet and glaciers released raging torrents of water that coursed across the landscape.

The final stages of the Ice Age in Scotland were complicated by events that happened across the Atlantic Ocean. As the climate started to improve at the end of the last advance of the ice, so Scotland was plunged back into the grip of winter. Although there is no universal agreement amongst academics on this, one version of the story is that a huge ice dam, holding back an enormous lake of meltwater that stretched across America, was eventually breached and water gushed into the North Atlantic in colossal quantities. This disrupted the Gulf

Gleneagles was carved through the flint-hard lavas of the Ochil Hills by a glacier. This demonstrates the huge erosive power of the ice in moulding the landscape.

Stream, a benign flow of warmer water that wafts northwards from the Gulf of Mexico and, in so doing, significantly improves our climate. With this moderating influence temporarily lost, the temperatures once again crashed and evidence exists for the formation of ice around in the west. This event is known as the Loch Lomond Readvance. The temperature balance was restored later, when the Gulf Stream re-established itself.

Temperatures took another nosedive with the onset of the Little Ice Age. The dates of this event are still debated, but between the sixteenth and nineteenth century is a reasonable estimate. It was not a full-blown ice age, so the name is slightly misleading, but winter temperatures periodically fell well below freezing. Life must have been uncomfortable for all who lived at these times. Frost fairs took place for many years in London on the frozen River Thames, and Scottish artist Henry Raeburn painted the Revered Robert Walker skating on an ice-covered Duddingston Loch. This work is dated at 1790.

There is little consensus on the cause of this worldwide event, but changes in solar radiation and the modification of the pattern of ocean current circulation are two possible culprits. This dramatic change in

'The Skating Minister' – Reverend Robert Walker (1755–1808) Skating on Duddingston Loch (1790s) by Henry Raeburn.

our past climate demonstrates just how sensitively balanced our natural atmospheric conditions are. We would be very well advised not to ignore the obvious signals of a looming climate catastrophe, which are so clearly emerging. This is no hoax.

After the ice

As the current phase of the Ice Age ended, huge quantities of water were liberated from the ice and sea levels rose worldwide to a much higher point than today. Another readjustment was also underway. The landmass of Scotland had been weighed down by ice and it rebounded by a few metres to a higher level when that burden was lifted. This process is called 'isostatic readjustment'. There was a complex interplay between higher sea levels and a rising landmass. The net result, around the coastline of Scotland, was that many areas formerly swept by the sea were lifted up to become dry land. We call these areas that were formerly intertidal 'raised beaches'.

The carse lands, overlooked by Stirling Castle, were once tidal mudflats and part of the Forth Estuary. After the ice melted, the

Below.
Turnberry Golf Course on the Ayrshire coast is built on a raised beach. This area was formerly swept by the waves when sea levels were higher.

coastline moved inland, driven by rising sea levels. The sediments that lie beneath the productive farmland have yielded the remains of marine animals. The most surprising find was whale bones! During the eighteenth and nineteenth centuries, when efforts were being made to 'improve' the land through drainage and removal of peat, twelve whale skeletons were discovered west of Stirling. A blue whale skeleton measuring 7 metres in length was found near the site of the University of Stirling.

The Forth Estuary formerly extended west beyond Stirling. It is now dry land used for farming and housing.

8
Golf – nature provides the perfect stage

Golf is our national game and has been played on the coastal links at St Andrews since the fifteenth century. It was mentioned in an Act of the Scottish Parliament in 1457 – at that time, citizens were forbidden from playing, as it interfered with their archery practice. This ban was later rescinded when King James IV himself became hooked on the game!

Lying just to the north of the ancient university town of St Andrews, the forces of nature have conspired to create the perfect pitch. Sand accumulated into dunes, blown inland from the beach on howling easterly winds, was then stabilised by marram grass. This dramatic landscape was fashioned by the elements into a series of ridges and hollows, a canvas for the first golf course anywhere in the world to be created in 1552.

In the earliest days, the stage for the first golfing contests was the untouched natural links land, created solely by the winds, waves and

The Old Course and the West Sands at St Andrews, as seen from the air. Long-shore-drift transported sand along the West Sands and, during periods of high wind and storms, the grains dumped in the intertidal area were blown on shore, forming impressive steep-sided dunes. These shifting landforms were naturally stabilised by the growth of marram grass, whose deep roots have 'frozen' the dunes in time.

tides. The contours of the land are still largely natural, although tees and greens were later added to allow competitive play. This championship course challenges the very best exponents of the game. It is regularly chosen as the venue for the Open Championship. The Old Course at St Andrews is universally regarded as the 'home of golf'. The sport is now played by over 25 million people worldwide, but all pay homage to this place as the hallowed ground where it all began.

Carnoustie, near Dundee, Muirfield in East Lothian, Kings Barns in Fife, Royal Troon and Turnberry on the Ayrshire coast are also world-renowned links venues for the game. Again, the natural elements of wind and sea take credit for the formation of this prime golfing real estate.

Gleneagles, in lowland Perthshire by Auchterarder, is another golfing gem. Here, nature also created the contours of the land to fashion the perfect golfing stage. But this time the forces at work were ice and water. During the last glaciation, ice was piled high over the landscape. When the snows melted, it was clear that it had created the perfect natural canvas for James Braid, the doyen of golf course designers, to work his magic. The fairways snake between ridges of sand and gravel called eskers. These sinuous features, some 10 metres high and up to a kilometre in length, were dumped by water flowing beneath the ice. Floodwaters were generated in prodigious quantities as the ice melted in response to rising global temperatures. These eskers ridges are the perfect viewing platform for the many golfing enthusiasts who flock to this place on match day. The Ryder Cup was contested here in 2014 and the Solheim Cup four years later. These global sporting events pit the best from Europe, men and women, against their counterparts from the USA. For the three days of both competitions, this exquisite natural landscape, fashioned by ice and water, was the centre of attention of the international sporting world.

This is the 18th green of the PGA Centenary course at Gleneagles, where the drama of the Solheim Cup competition reached a crescendo. The match was won by Europe, with the last putt of the deciding singles contest, with the world's sporting press in attendance. The natural amphitheatre around the green provided a fitting venue.

9
Places to visit

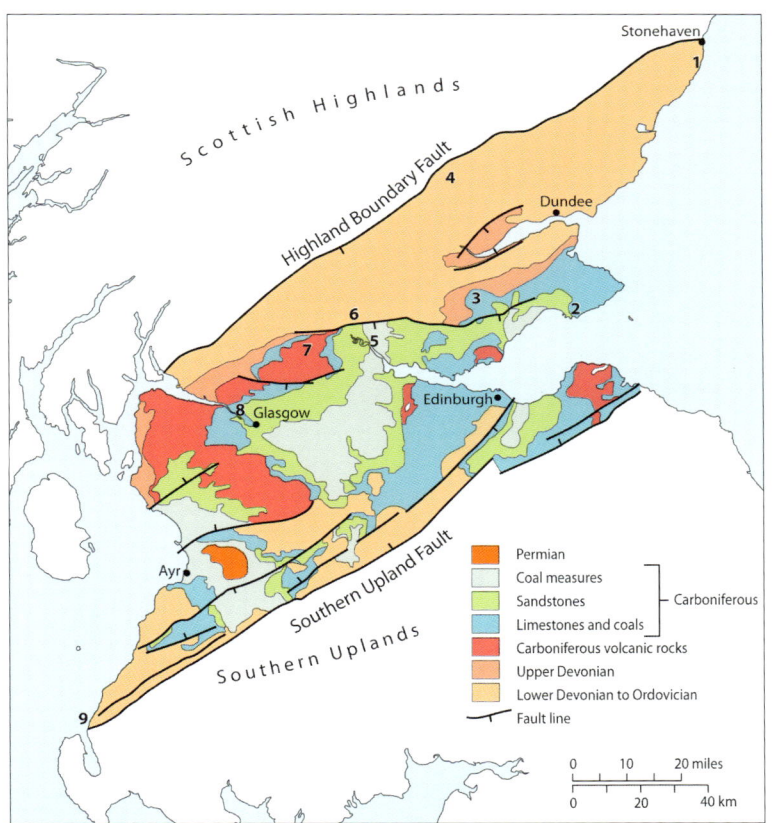

This short section gives a few suggestions of interesting places to visit for those who are unfamiliar with the area. It's not an exhaustive list, but it's a start. The area is covered by twelve 1:50,000 scale OS Landranger maps and the Bedrock Geology UK North map, published by the British Geological Survey. These maps will help you plan your visit and navigate around the sites. Some of these locations are described in greater detail in the foregoing text.

1. **The cliffs around Dunnottar Castle:** just to the south of Aberdeen. The cliffs show excellent exposure of Old Red Sandstone conglomerates and lava flows.

2. **The cliffs around Elie Ness:** found on the Fife coast, the cliffs provide exposure of the volcanic vents that tapped a deep source. Lumps of lower crust and mantle are embedded in the lavas that travelled around 70 km from the depths of the lower crust to the surface. Please don't hammer or collect lumps of rock, as this is an internationally important research site.

3. **The lower slopes of the Lomond Hills:** near Kinross and Loch Leven, this area comprises limestone and sandstones of Lower Carboniferous age. The upper part of these west-facing slopes is an impressive example of the Midland Valley Sill that is such an important landscape feature.

4. **Bennybeg:** south of Crieff, this site was visited by James Hutton en route to the Highlands. This wall of rock is a dyke, Carboniferous in age. Dykes are pulses of molten magma that cut across existing rock structures. This was one of the sites that Hutton used to develop his ideas that the Earth was a heat engine. The adjacent garden centre tolerates visitors to their car park, but repay them by buying some plants!

5. The Stirling Smith Art Gallery and Museum: displays of the whale bones found locally can been seen here. Also find out more about William Wallace, Roman pottery and the world's oldest football!

6. Wolf's Hole quarry (NS 7893 9808): long disused, it provided building stone for the expanding town of Stirling. The sandstones have yielded fish remains and were deposited in a braided river system. The sandstones are overlain by a lava flow related to the Ochil Hills Volcanics.

7. The Gargunnock Hills: built from layer upon layer of lavas, this site is one of the best to see the trap or stepped topography and also the lateral extent of individual lava flows. They can be viewed from the A811 that runs to the north of the Gargunnock Hills.

PLACES TO VISIT

8. Fossil Grove: located in Victoria Park in Glasgow. There are limited opening hours, but it's very much worth a visit. Fossil Grove has its own website.

9. The raised beach at Ballantrae was formed when sea levels were much higher than today. The foot of the ancient cliffline, now covered by gorse, was lapped by the sea around 7,000 years ago. The fields between the road and the ancient cliffline were part of the intertidal zone that lay between the low and high tide marks.

Acknowledgements and picture credits

Thanks are due to Professor Stuart Monro OBE FRSE and Moira McKirdy MBE for their comment and suggestions on the various drafts of this book. I also thank Debs Warner, Mairi Sutherland, Andrew Simmons and Hugh Andrew from Birlinn Ltd for their support and direction. Mark Blackadder's book design is up to his usual very high standard. I thank the James Hutton Foundation for their financial support in the publication of this book. Scottish Natural Heritage, in association with the British Geological Survey, published the *Landscape Fashioned by Geology* series that was the precursor to the new *Landscapes in Stone* titles. I thank them both for their permission to use some of the original artwork and photography in this book. Mike Browne, Alan McKirdy, David McAdam, John Gordon and Colin MacFadyen from BGS and SNH wrote the original texts. I dedicate this book to Percy Alan Klar, our firstborn grandchild. I had forgotten the boundless joy and endless fun that a new life bestows on the immediate and wider family. Good luck to the new parents, Fiona and Greg.

Picture credits

2–3 AlisterFirth/Alamy Stock Photo; 6 richardjohnson/Shutterstock; 10 Helen Stirling (map); 12 (upper) Granger, NYC/Alamy Stock Photo, (lower) drawn by Jim Lewis; 13 drawn by Jim Lewis; 14 (upper) © National Museums Scotland, (lower) used with permission, Nigel Trewin, *Scottish Fossils*; 15 courtesy of Professor Euan Clarkson; 16 (upper) Craig Ellery, (lower) Lorne Gill/SNH; 17 drawn by Jim Lewis; 18 (upper) Clare Hewitt, (lower) Lorne Gill/SNH; 19 Lorne Gill/SNH; 20 used with permission, Nigel Trewin, *Scottish Fossils*; 21 Stuart Allison, St Andrews University; 22 Lorne Gill/SNH; 24 (lower) Lorne Gill/SNH; 25 Lorne Gill/SNH; 26 (upper) Peter Gaeng/Shutterstock, (lower) Jackson-Carter/Shutterstock; 29 Lorne Gill/SNH; 30 Hunterian Museum and Art Gallery, University of Glasgow; 31 (upper) Clare Hewitt; 31 Patricia and Angus Macdonald/Aerographica; 32 drawn by Robert Nelmes; 34 Lorne Gill/SNH; 35 (upper) Lorne Gill/SNH, (lower) Alan McKirdy; 37 drawn by Jim Lewis; 38 Lorne Gill/SNH; 39 GL Archive/Alamy Stock Photo; 40 Lorne Gill/SNH; 41 Patricia and Angus Macdonald/Aerographica; 42 Alan McKirdy; 43 Helen Stirling (map); 44 Alan McKirdy; 45 Lorne Gill/SNH; 46 (upper) Mike Browne, (lower) Lorne Gill/SNH; 47 Patricia and Angus Macdonald/Aerographica.

Writing words with ll   Unit 2

1 Trace and write the letters.

ll
___

ll
___

ll
___

2 Trace and write the ll.

hill

hi___

sell

se___

doll

do___

3 Show if the words are things you see or things you do.

see   do

← doll

hill

sell

yell

bell

Tick your best letter.

Unit 3 P Introducing capitals for long-legged giraffe letters

1 Trace and write the capital letters.

2 Trace the letters. Join each lower case letter to its capital letter.

3 Write your name. Don't forget the capital letter.

Tick your best letter.

CAMBRIDGE

Penpals for Handwriting

1

Workbook

Name   *Class*

Unit 1 **P** **Practising long-legged giraffe letters**

1 Trace and write the letters.

l   u

i   j

t   y

2 Write the punctuation marks.

.☐   .☐   !☐   !☐

3 Finish the words. Write a punctuation mark in each box.

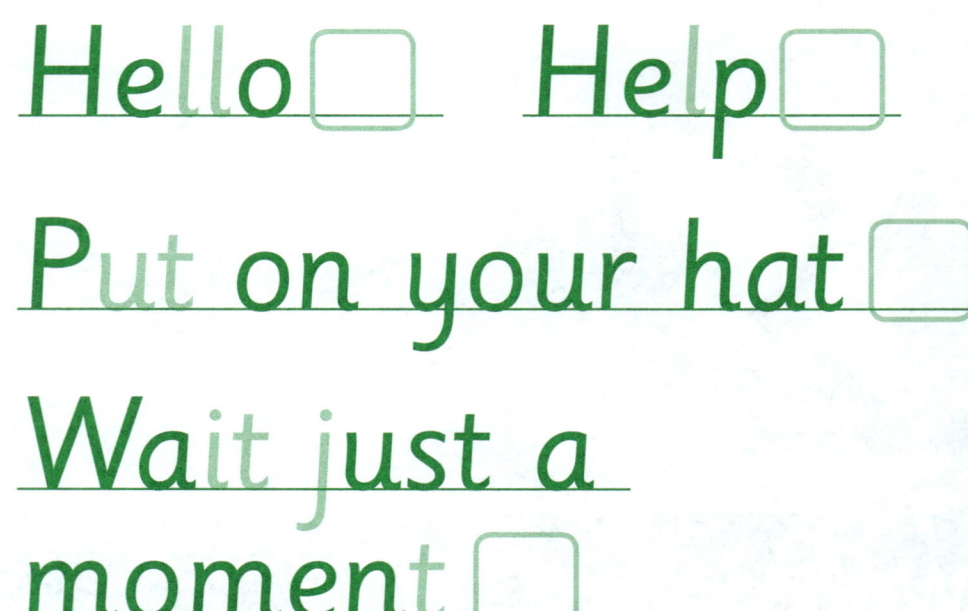

Hello☐   Help☐

Put on your hat☐

Wait just a moment☐

Tick your best letter. ✓

Practising one-armed robot letters G Unit 4

1 Trace and write the letters.

r _____   n _____   m _____   p _____
h _____   b _____   k _____

2 Write un before each word to make the opposite.

unpack

___fit

___plug

___lock

Tick your best letter. ✓

Unit 5 S — Practising long-legged giraffe letters and one-armed robot letters

1 Trace and write the letters.

b _____ p _____ y _____ t _____

2 Trace and write the words.

my _____   the _____   put _____

_____   _____   _____

3 Choose a word to fill each gap.

| my the put |

Can I _____ hat on _____ peg?

4 Write the three missing words again.

_____   _____   _____

Tick your best word.

Introducing capitals for one-armed robot letters  Unit 6

1 Trace and write the capital letters.

R ____   N ____   M ____   H ____

K ____   P ____   B ____

2 Trace the letters. Join each lower case letter to its capital letter.

3 Write your name. Don't forget the capital letter.

_____

Tick your best capital letter.

Unit 7 **Practising curly caterpillar letters**

1 Trace and write the letters.

c _____ a _____ d _____ o _____

s _____ g _____ e _____ f _____

2 Add s to show more than one. Rewrite each word as a plural.

One	More than one
cat	_____
dog	_____
goat	_____
tiger	_____

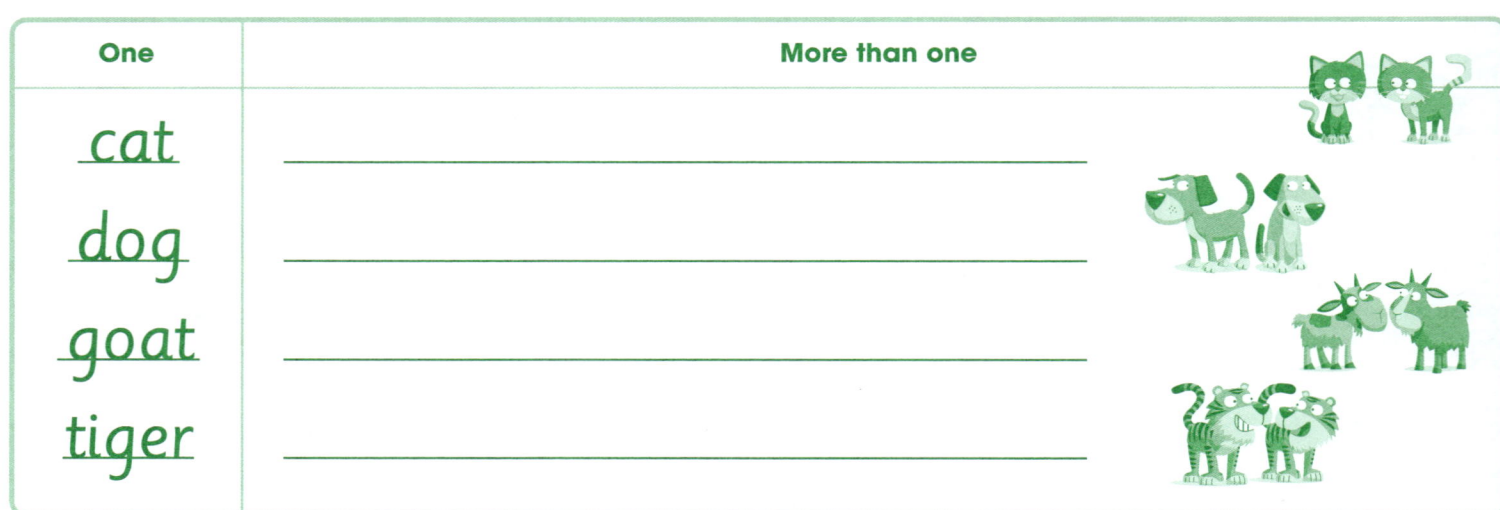

3 Write one more plural word.

_____

Tick your best letter.

8

Writing words with double ff　　　　　　　　　　　　　　Ⓟ Unit 8

1 Trace and write the letters.

f _____　　ff _____　　ff _____

2 Trace and write the words and punctuation mark.

of _____　　off _____　　? _____

_____　　_____　　_____

3 Write of **or** off **in each gap. Add** ? **or** . **in the boxes.**

Did you fall _____ your bike ☐

Can I have a cup _____ milk ☐

Turn _____ the light ☐

4 Write a sentence with the word off**.**

_____

Tick your best question mark. ✓

Unit 9 S

Writing words with double ss

1 Trace and write the letters.

ss

___

ss

___

ss

___

ss

___

2 Trace and write the words.

press

___

cross

___

lesson

___

unless

___

3 Say and clap the words. Show the number of claps or syllables.

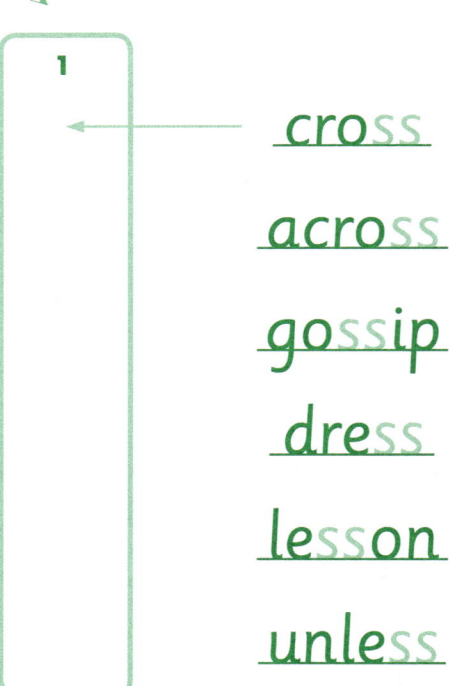

1 — cross

2 — across, gossip, dress, lesson, unless

4 Write another word with ss. It can rhyme with mess or miss.

___

Tick your best letter. ✓

10

Introducing capitals for curly caterpillar letters   P Unit 10

1 Trace and write the capital letters.

C   A

D   O

S   G

E   F

2 Write these signs in capital letters.

stop

way out

fire exit

push

3 Write another sign in capital letters.

Tick your best letter.

Unit 11 P   Practising long-legged giraffe letters, one-armed robot letters and curly caterpillar letters

1 Trace and write the words.

I _____   me _____   my _____

he _____   him _____   his _____

2 Write he, him **or** his **in the gap. Add** ? **or** . **in the boxes.**

This is Sanjay☐ I am _____ friend☐ Can I give _____ a bun because _____ gave me a cake☐

3 Tick words you would use if you were writing about a girl.

she ☑   him ☐   her ☐   it ☐

12

Practising zig-zag monster letters — Unit 12

1 Trace and write the letters.

v

w

x

z

2 Trace and write the words.

work

doze

box

cave

3 Join pairs of these words to make longer words.

wood   dozer

bull   work

match   man

cave   box

4 Write the words you made.

woodwork

Tick your best word.

Unit 13 G

Writing words with double zz

1 Trace and write the letters.

zz

zz

zz

2 Trace and write the words.

buzzer

pizza

puzzle

3 Add s or es to show more than one. Rewrite each word as a plural.

One	More than one
pizza	_____
puzzle	_____
buzz	_____
sizzle	_____
buzzer	_____

4 Write a sentence using one of the zz words.

_____

_____

_____

Tick your best letter. ☑

Mixing all the letter families

G Unit 14

1 Trace and write the letters.

ss

ss

ss

ss

2 Trace and write the words.

was

with

were

very

3 Write was, with, were, very in the gaps.

Will and Zak ____ in the ____ park☐ It ____ raining☐ Zak got ____ muddy ____ Will☐

4 Write a sentence using the word very.

_____

_____

Tick your best word. ☑

15

Unit 15 P   **Practising all the capital letters**

1 Trace and write the letters.

2 Trace and write the names.

3 Write a name for each of the children. Choose any name.

v V

Vikash

_____

_____

_____

_____

w W

Wayne

_____

_____

x X

Xavier

_____

4 Write the names of people in your family.

_____

_____

z Z

Zac

_____

_____

_____

_____

Tick your best name.

16

Practising all the numbers: 0–9　　　S　Unit 16

1 Trace and write the numbers.

0 ___　1 ___

2 ___　3 ___

4 ___　5 ___

6 ___　7 ___

8 ___　9 ___

2 Read and write the number words.

five _____　four _____

three _____　six _____

nine _____　zero _____

one _____　eight _____

seven _____　two _____

3 Write the number words from 0–9 in order.

_____

_____

Tick your best number.

17

Unit 17 G **Writing words with** ck **and** qu

1 Trace and write the letters.

qu ____   qu ____   ck ____   ck ____

2 Trace and write the words.

quack   squirrel   quick

_____   _____   _____

3 Which words can you make plural? Write s or es at the end of those words.

chicken ___   brick ___

queen ___   clock ___

buckle ___   quiet ___

4 Write three more words that have ck or qu in them.

_____

_____

Tick your best word. ✓

Practising long vowel phonemes: *ai, igh, oo*

Unit 18

1 Trace and write the letters.

ai

igh

oo

2 Trace and write the words.

faint

high

smooth

3 Add *-er* and *-est* to each word.

	+ er	+ est
cool	cooler	_____
faint	_____	_____
smooth	_____	_____
high	_____	_____
bright	_____	_____

4 Write a sentence using an *-est* word.

Tick your best word.

Unit 19 G Practising vowels with adjacent consonants: ee, oa, oo

1 Trace and write the letters.

2 Trace and write the words.

3 Add -ed and -ing to each word.

ee   needed

	+ ed	+ ing
need	needed	needing
peel		
groan		
cook		

oa   moaned

4 Use an -ed word in a sentence.

oo   looked

Tick your best word. ✓

End-of-term check Unit 20

1 Trace the days.

Saturday
Tuesday
Sunday
Thursday
Monday
Wednesday
Friday

2 Write the days in order.

Tick your best word.

Unit 21 Ⓢ **Numbers** *10–20*: spacing

1 Trace and write the numbers.

10   11   12   13   14   15
16   17   18   19   20

2 Write the numbers beside the words.

seventeen ____   twenty ____
ten ____   thirteen ____   fifteen ____
fourteen ____   nineteen ____
eleven ____   twelve ____
sixteen ____   eighteen ____

3 Practise writing two of the number words.

Tick your best number.

Practising *ch* unjoined

Unit 22

1 Trace and write the letters.

ch _____   ch _____   ch _____   ch _____

2 Trace and write the words.

chips   cheese   chicken   lunch

_____   _____   _____   _____

3 Choose *ch* words to write in the gaps. Write . or ? in the boxes.

What will we have for _____ today ☐

I want _____ ☐

He wants _____ ☐

She wants _____ ☐

Tick your best letter. ☑

Unit 23 G Introducing diagonal join to ascender: *ch*

1 Trace and write the letters.

2 Trace and write the words.

3 Rewrite each word. Join *ch* and add *-ing*.

4 Practise joining *ch* in one more word.

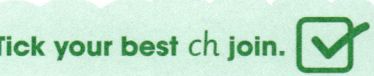

Words with ai unjoined — Unit 24

1 Trace and write the letters.

ai ____ ai ____ ai ____ ai ____

2 Trace and write the words.

rain   train   ail   tail

_____   _____   _____   _____

3 Make sets of rhyming words.

train   sp_____   n_____   t_____

           (rain)   (ail)

b_____   g_____   m_____

4 Think of another rhyming word for each.

_____   _____

Tick your best word.

Unit 25 S Introducing diagonal join, no ascender: *ai*

1 Trace and write the letters.

ai _____   ai _____   ai _____   ai _____

2 Trace and write the words.

rain   rail   paint   paint

____   ____   ____   ____

3 Do the word sum. Write the word. Join *ai*.

rain + bow = rainbow

rail + way = _____

paint + brush = _____

finger + nail = _____

4 Write all the long words you made again with *ai* joined.

_____   _____

_____   _____

Tick your best join. ✓

Practising wh unjoined  Unit 26

1 Trace and write the letters.

wh _____ wh _____ wh _____ wh _____

2 Trace and write the words.

when _____ what _____ where _____

3 Write a wh word at the beginning of each sentence.

_____ is your house ☐

_____ will it be my birthday ☐

_____ can you see ☐

4 Write another word beginning with wh.

_____

Unit 27 G P   Introducing horizontal join to ascender: *wh*

1 Trace and write the letters.

wh _____ *wh* _____ *wh* _____ *wh* _____

2 Trace and write the words.

whale   *whale*   *white*   *white*

_____ _____ _____ _____

3 Put the words in order to make a sentence.

Can the _____

_____ _____ along ☐

| whale |
| whizz |
| white |

Add the punctuation in the box. Tick the punctuation you used.

4 Practice writing *wh* with joins.

_____

Tick your best *wh* join. ✓

Practising ow unjoined   G Unit 28

1 Trace and write the letters.

ow _____   ow _____   ow _____   ow _____

2 Trace and write the words.

owl   cow   bowl

_____   _____   _____

3 Rewrite each word as a plural. Add s to show more than one.

One	Two
owl	_____
cow	_____

  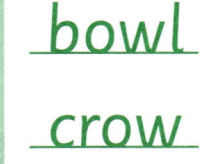

One	Two
bowl	_____
crow	_____

4 Write more plural words.

_____

_____

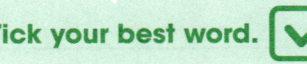
Tick your best word.

Unit 29   Introducing horizontal join, no ascender: ow

1 Trace and write the letters.

ow ____   ow ____   ow ____   ow ____

2 Trace and write the words.

show   show   show   show

_____   _____   _____   _____

3 Ring words like show in yellow and words like crown in brown. Copy the words.

show _____   crown _____

how _____   slow _____   owl _____

4 Write one more word that rhymes with show. Write one more word that rhymes with crown.

_____

_____

Tick your best ow join.

Assessment   P G Unit 30

1 Trace and write the letters.

ch _____ ai _____ wh _____ ow _____

2 Trace and write the words.

reach   wait   when   rainbow

_____ _____ _____ _____

3 Use *reach, wait, when* or *rainbow* to fill in the gaps. Write . ? or ! in the boxes.

Can you tell me _____ ☐

I can see a _____ ☐

I can try to _____ it ☐

I can't _____ ☐

Tick the punctuation you used. . ☐   ? ☐   ! ☐

Certificate

for completing

PENPALS for
Handwriting 1

awarded to

NAME
_____

DATE   SIGNED
_____   _____

Shaftesbury Road, Cambridge CB2 8EA,
United Kingdom

One Liberty Plaza, 20th Floor, New York,
NY 10006, USA

477 Williamstown Road, Port Melbourne,
VIC 3207, Australia

314–321, 3rd Floor, Plot 3, Splendor Forum, Jasola
District Centre, New Delhi – 110025, India

103 Penang Road, #05-06/07, Visioncrest Commercial,
Singapore 238467

Cambridge University Press is part of the
University of Cambridge.

It furthers the University's mission by
disseminating knowledge in the pursuit of
education, learning and research at the
highest international levels of excellence.

Information on this title: www.cambridge.org

© Cambridge University Press & Assessment 2015

This publication is in copyright. Subject to statutory
exception and to the provisions of relevant collective
licensing agreements, no reproduction of any part may
take place without the written permission of Cambridge
University Press.

First published 2015

20 19 18 17 16 15 14

Printed in Great Britain by Ashford Colour Ltd.

*A catalogue record for this publication
is available from the British Library*

ISBN 978-1-84565-440-5

Acknowledgements

© Cambridge University Press & Assessment 2015
www.cambridge.org

Illustrations by Marek Jagucki
Cover design and layout by me&him
Authors: Gill Budgell and Kate Ruttle

CAMBRIDGE
UNIVERSITY PRESS

www.cambridge.org